中国风筝节
Chinese Kite Festival

Rich Lo

Holiday House New York

A fish leaps into the blue sky.
鱼跃入蓝天。

A crab moves sideways
beneath the clouds.
螃蟹在云层下横行移动。

A dragonfly hovers in the breeze.
蜻蜓在微风中盘旋。

A bird soars from its nest.
鸟从鸟巢中展翅上腾。

A butterfly flutters in the wind.
蝴蝶在风中飘扬。

A tiger pounces across the sky.
老虎在天空中猛扑。

A ladybug takes flight from a leaf.
瓢虫从叶子上起飞。

A frog jumps into the air.
青蛙跳入空中。

A bee flies above a sunflower field.
蜜蜂在向日葵田上飞翔。

A **bat** floats across the moon.

蝙蝠飘过月亮。

A turtle slides from a rock.

乌龟从一块岩石滑行。

A goldfish swims in the rain.
金鱼在雨中游玩。

A rooster struts with the current.
公鸡随着气流昂首阔步。

An owl hoots above the trees.
猫头鹰在树上鸣叫。

ANIMALS in Chinese Culture

CRAB
Crabs signify prosperity and success.

TIGER
Tigers represent strength and courage.

DRAGONFLY
Dragonflies represent harmony and prosperity.

BUTTERFLY
Butterflies are a sign of love.

FROG
Frogs are a symbol of healing and prosperity.

FISH
Fish are a symbol of abundance and prosperity.

OWL
Owls personify bravery.

BIRD

Birds are associated with the sun and vitality.

BAT

Bats represent happiness and wealth.

BEE

Bees signify hard work.

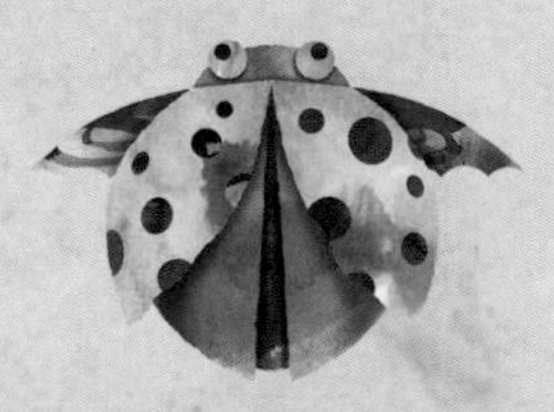

LADYBUG

Ladybugs bring good luck.

GOLDFISH

Goldfish stand for wealth and good fortune.

TURTLE

Turtles symbolize long life and immortality.

ROOSTER

Roosters represent reliability and honesty.

DEDICATED TO MY PARENTS AND SISTER ALICE.

The publisher wishes to thank Belinda Chan for her expert translation of the text.

Printed and bound in August 2021 at C&C Offset, Shenzhen, China.
www.holidayhouse.com
First Edition
1 3 5 7 9 10 8 6 4 2

Library of Congress Cataloging-in-Publication Data

Names: Lo, Rich, author, illustrator.
Title: Chinese kite festival / Richard Lo.
Description: First edition. | New York : Holiday House, 2021. | Audience: Ages 4–8. | Audience: Grades K–1. | Summary: "Animal names and their significance in Chinese culture is explained in simple bilingual text for young readers"—Provided by publisher.
Identifiers: LCCN 2021010288 | ISBN 9780823447640 (hardcover)
Subjects: LCSH: Kites—China—Juvenile literature.
Festivals—China—Juvenile literature. | Holidays—China—Juvenile literature. | Picture books for children.
China—Civilization—Juvenile literature. | LCGFT: Picture books.
Classification: LCC QL737.C25 L62 2021 | DDC 796.15/80951—dc23
LC record available at https://lccn.loc.gov/2021010288

ISBN: 978-0-8234-4764-0 (hardcover)

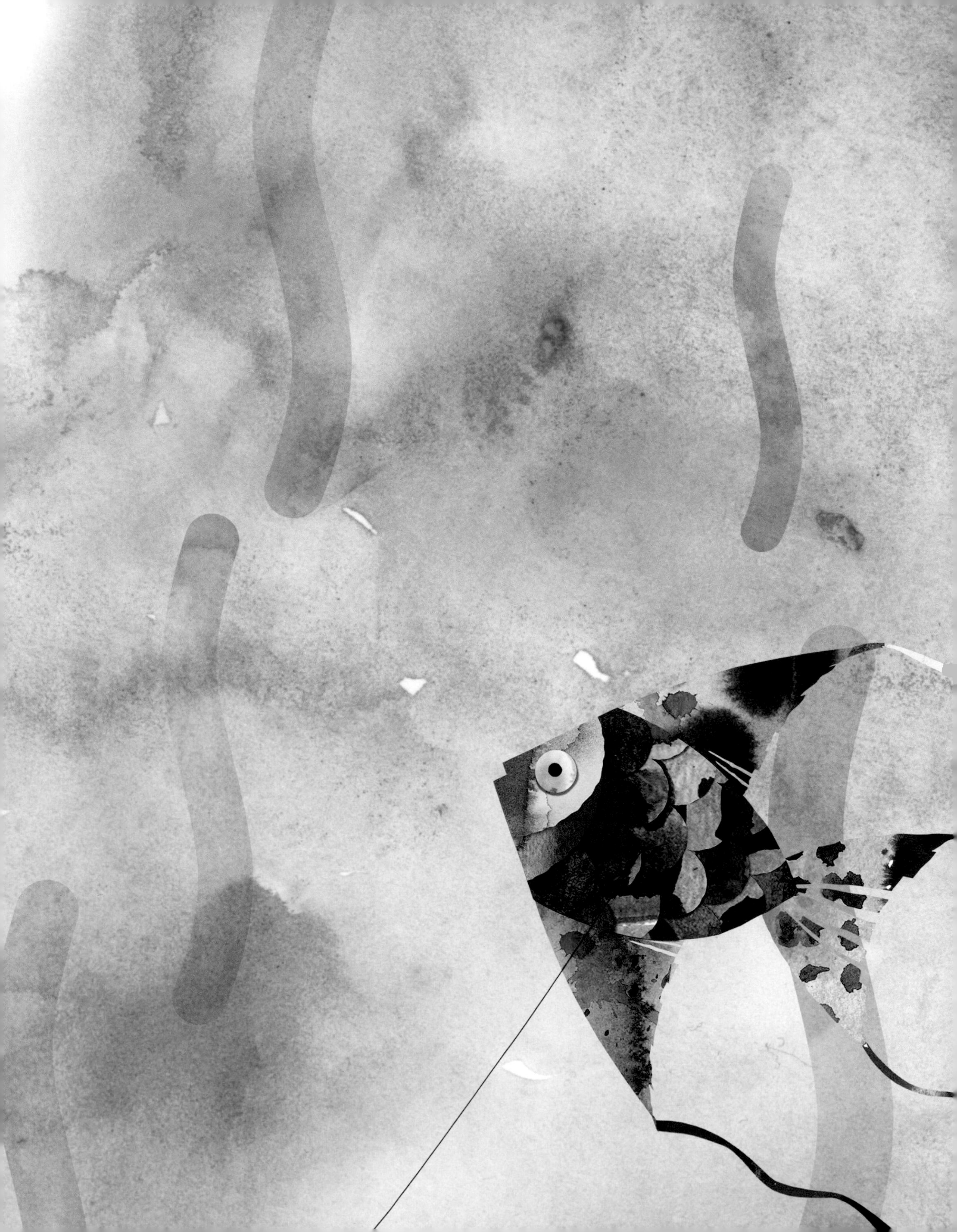